THE VELOCITY OF LIGHT AND NUCLEAR FORCE

It is well established that the forces between nucleons are transmitted by meson. The quantitative explanation of nuclear forces in terms of meson theory was extremely tentative & incomplete, but this theory supplies a valuable point of view. It is fairly certain now that the nucleons within nuclear matter are in a state made rather different from their free condition by the proximity of other nucleons.

Now we have the rest mass energy $= m_0 c^2$

Differentiating with respect to r (Inner radius at which nuclear force comes into play)

$$\frac{d(m_0 c^2)}{dr} = c^2 \frac{dm_0}{dr} + m_0 \frac{d(c^2)}{dr} = c^2 \frac{dm_0}{dr} + m_0 \frac{d(c^2)}{dc} \frac{dc}{dr} = c^2 \frac{dm_0}{dr} + 2m_0 c \cdot \frac{dc}{dr}$$

. This force is short range, attractive & along the line joining the two particles (central force).(The wide success of this first application of quantum mechanics to nuclear phenomena gives us confidence in general use of quantum mechanics for the description of the force between heavy particles in nuclei.

Where dm_0c^2 = energy of π^0 mesons or (k^0-meson)

dm_0 = mass of π^0 mesons or (k^0-meson)

m_0 = mass of nucleons

m_0cdc = energy of nucleons

dr= Range of nuclear force, which can be calculated from differentiation of Nuclear radius.(The force between two nucleons is attractive for distance r(radius) greater than dr (range) & is repulsive otherwise).This strongly suggests & well proved that to some degree of approximation the total isotopic spin T is a constant of the motion & is conserved in all processes, at least with a high probability.

dc= The average velocity of neutron &

proton. A large velocity is used in nuclear disintegration.

c = Velocity of light
2 = multiplicity of interacting particle is given by (2T+1), the isotopic spin has no such meaning for leptons or a gamma rays
1 = multiplicity of π^0 mesons (evidence of involving of neutral mesons)
Where T = Vector sum of isotopic spin of proton & proton & neutron & neutron. The success of these applications supplies additional support for the hypothesis of the charge independence of nuclear force. As the nuclear interactions do not extend to very large distances beyond the nuclear radius & this character is useful to solve the problem. The Full charge independence for any system in which the number of neutrons equals the numbers of protons, this formula give the evidence the

charge symmetry, merely means that the neutron-neutron & proton proton interaction are equal but says nothing about the relations of neutron proton interaction to others. Nuclear forces are symmetrical in neutrons & protons. i.e. the force between two protons are the same as those between two neutrons. This identity refers to the magnitude as well as the spin dependence of the forces.

The nucleus is a tightly bound system of the nucleons with a large potential energy as well as kinetic energy.(we can find it from observation)

Now we can see the following reaction

$P + P \text{-----} P + P + \pi^0$

$P + N \text{-----} P + N + \pi^0$

$\gamma + P \text{-----} P + \pi^0$

These reactions are soon as $Y + P \quad P + \pi^0$

(A similar qualitative, or in this case semi qualitative, understanding is gained of the interaction of the meson them selves with nucleons, as manifests by their scattering from nucleons & their production from nucleons by Gamma rays at moderate energies.

$$P + P \; \text{---} \; P + P + \pi^0$$
$$P + N \; \text{-----} \; P + N + \pi^0$$

Charge independence of nuclear forces demand the existence of π^0 meson as amongst the same type of nucleons (p-p) or (N-N). This force demand the same spin & orbital angular momentum. Positive pions are not able to surmount the nuclear coulomb barrier & there fore undergo spontaneous decay while negative pions are captured by nuclei. The exchange of a pion is thus equivalent to charge exchange. we can think of nucleons as exchanging their space & spin co-ordinates The exchange interaction is produced by only a neutral meson. The involving mesons without electric charge, that it gives exchange forces between proton & neutron & also therefore maintains charge independence character. In the neutral theory, therefore neutron & protons are completely equivalent & indistinguishable as far as the associated meson fields are concerned. Such particle decay into two gamma rays. These gamma rays are π^0 - rest systems are emitted in opposite direction & therefore spin π^0 must be Zero as the spin of photon is unity. It is evident from the nature of the products that neutral mesons decay by the electromagnetic interaction while charged pions decay by strong & weak interaction both. It means that neutral mesons constituents responsible for the electromagnetic interaction. Dramatically neutral meson plays the important role for the electro magnetic & nuclear force both. We know that neutron & proton can change into one another by meson capture. Protons &neutron can transform into each other by capture of positive & negative pion respectively, or get transform into the same particle through neutral meson interaction. It means that If the proton or neutron transform itself each other by charge meson, the nuclear force does not exist. But the pair of n-p has ability to produce neutral meson then the force must exist. During these transformation either an emission or an absorbtion of meson is essential. The attraction between any nucleons can arise from the transfer of a neutral meson from one nucleon to the other. If the meson were assumed to be charged (positive or negative) the resulting force between nuclear particles turned out to be of the exchange type which had been successful in the interpretation in nuclear physics. The mesons must obey Bose statistics because they are emmited in the transformation of a neutron into a proton(or vise versa) both obey Fermi statistics.

Important point--

1. The deuteron does posses measurable properties which might serve as a guide in the search for the correct nuclear interaction. The mass number of deuteron A is very minimum. From the findings we must regard the deuteron as loosely bound. The deuteron consist of two particle roughly equal mass M, so that the reduced mass of the system is $1/2M$. The deuteron has spin T=1, the neutron & protons spins might be a parallel combination. The magnetic moment of deuteron will, therefore be sum of magnetic moment of proton & neutron. According to conclusion, As the range becomes larger in deuteron nucleus & they becomes more unstable. The distance travel by neutral meson is twice of the normal range dr. The life time of these particle(mesons) is very large. If the average velocity of nucleons in reduced mass system are twice, then the force work properly. If the rest mass system becomes twice then this system will maintain stability. In whole

phenomena the separation of nucleons does not effected.

*When light nuclei of hydrogen atom comes within the range of nuclear force they can fuse together to form helium nucleus. In this process (fusion) the range is not effected. However the force is twice of the hydrogen's nuclear force & so on. Further the energy requires to bring nucleons inside the range of force is twice of the rest energy of hydrogen nucleus. In other word, we can say that the minimum energy requires to form a helium nucleus is twice of the hydrogen rest mass energy. The mass of He atom (alpha-particle) equal to the four times of the hydrogen atom, so that the nuclear force of helium nucleus is four times stronger than that of hydrogen atom. It means that the He atom is more stable than hydrogen atom. Becouse binding energy of helium atom is larger than that of hydrogen atom, so in this process large energy is released.(here the negative pion has significant role to produce nuclear interaction. While charge pion main tain the character of proton.) Now, as the hydrogen nucleus conveted into helium nucleus, These are happen when resonances of nucleons is in excited state. The scattering cross-section can be interpreted by assuming that in the strong interactions the total isospin is converted as well as the third component. The total isotopic spin of the system is 3/2. The ratio of radius & range is about 2:1. It shows that those nuclei has maximum number of nucleons are most stable than less mass number nuclei.

❖

2*. Also, Strongly, The velocity of light depends upon range of the nuclear force. The velocity of light equal like photons, lack mass & force carrying particle of other forces like strong force. Because range is variable, then the velocity of light must be variable. As, velocity of light= Range of nuclear force(distance travel by meson)/ time taken by meson. In this relation we can see that the velocity of light must be variable. Also, It is a function of energy. It is clear that the fundamental particles are not wholly independent, The neutron observed to change spontaneously into a proton. Neutron decay takes on the average some thousands seconds for free neutron, whereas within a nucleus the characteristic time between nucleon- nucleon collisions is 10^{-24} seconds. For a satisfactory picture it is often enough to think of the nucleus as a grouping of protons & neutrons interaction, with the appearance or disappearance of photons. One should be noted that this relation hold only inside the nucleus. Out side the nucleus the evident is to be contrary. It is a fact no body(even mesons or gamma rays) can have velocity greater than the velocity of light. From this formula we can find the nuclear force acts between the pair of nucleons & does not influenced by the presence of neighbouring nucleons. When we come to consider systems comprising more than two particles, we must expect some complication to enter. It is nessesary that any one particle must brings the velocity of light. We know that the nuclear force is short ranged. Out side of the range it is repulsive.

❖

*Range of nuclear force :- To show that the range of force is related to the mass of exchanged particle, assume that the $\pi 0$-meson is contained virtually in a proton. If this virtual particle travels with the velocity light as might be expected for a field particle, then greatest distance the meson could travel in this time also known as range of the pion exchange force.

3. It would seem that in a nucleus consisting of the many nucleons the binding energy per nucleon should increase with the increase of the mass number A . In reality evidence is to contrary, the binding energy per nucleons decreases with increasing mass number A, The binding energies of the different nucleon placed at various depths are not identical but depend upon the states of their actual binding in the potential well.The binding energies of the different nucleon placed at various depths are not identical but depends upon the state of their actual binding in the potential well. The range also depends upon mass number A & binding energy.weknow that the atomic mass number A is approximately equal to a twice the

atomic number Z. For the light & intermediate nuclei. It shows that light nuclei prefer to add nucleons is n-p pair. i.e there is a strong interaction between neutrons & protons. The range of nucalear force depends on the mass number A & the velocity of light depends on the range, so it is obviously thought that the spin & velocity of light depends on the mass number A of the nucleus & spin is zero or an integer for A even & is an odd half integral for A odd. The total rest mass energy also depends on the mass number A. For increasing of rest energy, we must increase the mass number A. obviously, the rest mass energy must be depends on the radial distance.

.The binding energy depends on the potential energy as well as kinetic energy, With the increasing kinetic energy of nucleons, the binding energy decreases & thus the nuclear force is stronger. The binding energy can be affected by kinetic energy as well as potential energy . The rest mass energy also depends on the kinetic energy of the particle. If the kinetic energy increases, then the rest mass energy decreases.The(N-Z) excess neutrons will have a much smaller binding. This is purely a quantum mechanical effect.If the mass number A increases the range decreases, & the force are stronger. This binding energy displays saturation effect. This property of the nuclear force can be explain in term of exchange nature of nuclear force. It should be noted that nucleons attract each other strongly only if they are in same orbital state. This formula prove the pauli hypothesis.

4.The velocity of light depends on the wavelength of it constituents,If the particles has longer wavelength then the range decreases & therefore force is stronger. we can find the effective range of nuclear force in terms of the Compton wave length of pi−meson. We know that different(variable) constituents (color particles) has different wavelength, so it is obviously thought the velocity light must be variable. Also, as we know, the density of all nuclei are equal thus the velocity of light must be equal for all nuclei. The forces responsible for binding the individual particle inside the nucleus must therefore be exceptionally strong. If the particles has motion then the material body has physical significance otherwise not. It means the force between elementary particles depends on the velocity of the body as well as mass of body. It should be remarked that the particles travels with velocity of light are not a conservable quantities. In this formula, we can observe that the photons or gamma rays are not mass less particles.

5. The striking correlation between mass of nucleons & number of nucleons suggest that the mass of nucleus is variable. It is a fact that the numbers are known by mass & velocity etc. without it, number has no meaning. Let a force acting upon a body displaces it by a distance dr then increase in kinetic energy of the body dT must be corresponding to the Work done is equal to $c^2 dm_0$ + 2 $m_0 c.dc$ The work done depends on the total meson productions. The total work done of nuclear particle depends on the range of nuclear force. This is increasing with increasing the range & decreases with decreasing the range. It should also be noted that the work done by fundamental particle depends on the mesons production. It is increasing with increase, similarly decreases with decreasing mesons productions.

It is evident that the kinetic energy of a body C^2 times the increase in the mass of the body from the rest mass

One should be noted that the kinetic energy of the body is independent of isotopic spin. The kinetic energy of the body can effect the interaction(Its effect the range) , How ever, total work done depends on the spin. An isotopic spin space can be defined just like the intrinsic spin in which strong interaction will be invariant under a rotation or in other words the isotopic spin will be a conservable quantity.

6 .There is no conservation law controlling the total number of K−aons or meson. The energy of formation of mesons comes from binding potential (which has the energy to formation of meson for a long time), but when these potential has not enough energy, the production of pions end & nuclear force does not exist.

7. A large fraction of total energy is oftenly interchanged between rest energy associated with mass & kinetic or potential energy. The sum of these three, the total energy is always conserved in any reaction, for example the rest energy of one k^0 is not great enough to make four(all type)poins . even if they could be make at rest.(but it is possible in $2k^0$). However the reaction, $k^0 \to \pi^{0+} \pi^0$ is allowed. Assume that the negative pion is captured at rest, the neutral meson & N equire a relative kinetic energy equal to given by energy mass relation.

K-- mesons are produced when high energy protons bombarded a suitable target. The pions are also produced in this process. With increasing proton energy the fraction of K—mesons relative to pions increases.

8. The energy require to pull out the nucleons from the nucleus is less than half of rest mass energy. The slow

motion neutron play this role. Similarly if the nucleus brings (from binding potential) sufficient energy for the existing of nuclear force. It maintains stability. In order to approach particle to within short range or closer the energy of the approaching particle should be very high.

9. It should be noted that the negative meson & positive meson has different masses, so that its rest energy are different. They have no role in charge independent nuclear interaction. They maintain the character of nucleons. The charge mesons plays no role in charge independence nuclear force. The emission of a charged meson will be accompanied by a change of chatge of the emitting nuclear particle, Thus a neutron can only emit a negative meson or absorb a positive meson & will thereby be

transformed into proton. When we consider the emission of one meson by a nuclear particle & reabsorbtion by another. This will lead to forces between a neutron & a proton. The negative & positive charge meson comes to close together, the can neutralize each other then the force between neutron & proton come into play. So obviously we can say that only neutral meson plays important role in charge independent nuclear force. The mesons(positive & negative) can be absorbed by the nucleus of an element or it may be combine with the another meson then the sum of the masses of these mesons converted into energy. This process is called annihilation of matter. Before this process one positive meson & one negative meson unite to make neutral particles called k^0 meson. The process of construction & destruction has proved very help full is considering the origin of universe. We know that neutral—meson decay into two photons & never into three photons. This implies that the intrinsic spin must be zero. Thus we see that all pions have spin zero & have no magnetic moment. It is clear that the neutral pions has been produced by bombarding hydrogen & deuteron with high energy photons. Gamma rays has sufficient energy to maintain energy of nucleons then the nucleons produce the neutral mesons. One can speak of the meson field associated with a proton(or a netron) because the nature (charge) of the nuclear particle does not change by emitting or absorbing a neutral meson. It has developed a theory of nuclear forces in which neutral way the equality of the forces between like & unlike nuclear particles. An alternative way of explaining this equality is to assume interaction with neutral meson only. Then the charge of the nuclear particles(whether it is a proton or a neutron)

becomes entirely irrelevant & the equality of forces follows immediately. This alternative is discussed in present paper.
9. According to the pauli principle only two neutrons & two protons will found in the same orbital state. Therefore it is possible to find four nucleons strongly bound or Alpha particle structure, also confirm by binding energy curve. The extraordinary stability of the alpha shows that the most stable nuclei are those in which number of nucleons & photons are equal. We can find it from this formula. It is obviously thought that the full charge independence for any system in which the number of neutrons equal to the number of protons. From conclusion, we get, Number of photons = number of nucleons= 2(number of neutral mesons)

10. Due to the rapid decline of nuclear force with distance, a positively charge particle will experience diminishing attraction near the surface of the nucleus when receding from latter & at a certain distance equal to the nucleus r, the force of attraction will be balanced by the coulomb force of repulsion. The strong interaction is a short range force, conserves Baryon number B, charge Q, hypercharge Y, parity, isospin T, & its components. It is responsible for K-oans production, However, the decay of mesons, nucleaons & hyperons proceeded by an electromagnetic or weak interaction.

Broad Appeal – Since there is no requirement for the conservation of Pions so there is no conservation law in rest mass energy & even in the universe. This formula show that there is no meaning of the word' constant'

*The life time of any radioactive substances depends on the total number of pions production and another particles production. pions are commonly formed in the decay of k—oans, hyperons & resonant states. It should be noted that pions are formed only at high energy. Becouse of their short life time of neutral mesons move only a few atomic diameters before they decay(so that it influenced few neighbors nucleons)& thus are not affected by the matter through which they pass & thus nuclear force work properly. It should be also noted that in whole universe there is only mass will be conserved and energy will be destroyed, then the mass will not change into energy.

*It is enough to think that π- mesons which form a nuclear cloud around the individual nucleons & are in a virtual state get their requisite rest mass energy from the incident particle & are released from the nuclear binding potential. Nuclear binding potential compensate the rest mass energy. It produce enough energy to maintain the rest mass energy for production of mesons. Since the rest mass energy of π - mesons about 275 Me, the threshold energy for a gamma rays to produce the rest mass energy of these particles should be high. But, if protons projectiles are used to produce mesons, it requires a large threshold as a particle with mass retains some energy in the collision.

❖ Since there is no limitation of formation of Mesons even in strong interaction. This is due to high energy photons (γ - rays) then this cyclic chain should be possible

$\pi^0 \to \gamma$ rays

$\gamma + d \to d + \pi^0$

This reaction shows that the kinetic energy as well as potential energy of nucleon in the nucleus will be over and above of the rest mass energy. In these phenomena the total charge of fundamental particles are conserved.

❖ It is reasonable to assume that the nuclear force between two protons has the same characteristic as that between neutron & proton. The argument about short range forces involves both proton-proton & neutron - proton forces. The main difference between proton & neutron seems to be the electric charge, & the nuclear force apparently does not arise from charge. We assume there fore that the potential between two protons is confined within some short range as before, although the value of range need not necessarily be the same.

❖ The force depends on the separation of the nucleons not on the relative velocity or orientation of the nucleon spins with respect to the line. It should be noted that the force fully depends on the range of force.

❖ The core---- The interaction of the nucleons with havier mesons (e.g. the K-meson which have a mass of about 1000 electron masses) will not have much influence on the nuclear forces at low energy. (The repulsive core is not caused by the interaction of the nucleon with these havier meson, as was at one time suggested, but comes out naturally from its interaction with the ordinary pi-- meson). This conclusion is very important. Of course, when two nucleon collideat very high energy(billions of electron volts) all these argument break down. Then nucleon can pentrate or disturb each other core. This manifests by the multiple production of pi—meson & the frequent production of heavy meson. Encouraged by general argument that only the exchange of a relatively small number of mesons can have an appreciable effect on nuclear force at low & medium energies. We can find all these arguments from this formula.

- ❖ Resonances of nucleons-- Nucleon resonances or excited states, several resonant states of hyperons & mesons are produced for short time & then decay through strong interactions, In those cases where resonance widths are small, the strong force less stable. From observation we conclude that change of hydrogen nucleus into helium nucleus is due to excited state of nucleon.

www.ingramcontent.com/pod-product-compliance
Lightning Source LLC
Chambersburg PA
CBHW041120180526
45172CB00001B/343